25 Ultimate experiences

Islands

Make the most of your time on Earth

ROUGH GUIDES

25 YEARS 1982–2007

NEW YORK • LONDON • DELHI

Contents

Introduction

EXPERIENCES have always been at the heart of the Rough Guide concept. A group of us began writing the books **25 years ago** (hence this celebratory mini series) and wanted to share the kind of travels we had been doing ourselves. It seems bizarre to recall that in the early 1980s, travel was very much a minority pursuit. Sure, there was a lot of tourism around, and that was reflected in the guidebooks in print, which traipsed around the established sights with scarcely a backward look at the local population and their life. We wanted to change all that: to put a country or a city's popular culture centre stage, to highlight the clubs where you could hear local music, drink with people you hadn't come on holiday with, watch the local football, join in with the festivals. And of course we wanted to push travel a bit further, inspire readers with the confidence and knowledge to break away from established routes, to find pleasure and excitement in remote islands, or desert routes, or mountain treks, or in street culture.

Twenty-five years on, that thinking seems pretty obvious: we all want to experience something real about a destination, and to seek out travel's **ultimate experiences**. Which is exactly where these **25 books** come in. They are not in any sense a new series of guidebooks. We're happy with the series that we already have in print. Instead, the **25s** are a collection of ideas, enthusiasms and inspirations: a selection of the very best things to see or do – and not just before you die, but now. Each selection is gold dust. That's the brief to our writers: there is no room here for the average, no space fillers. Pick any one of our selections and you will enrich your travelling life.

But first of all, take the time to browse. Grab a half dozen of these books and let the ideas percolate … and then begin making your plans.

Mark Ellingham
Founder & Series Editor, Rough Guides

Ultimate
experiences
Islands

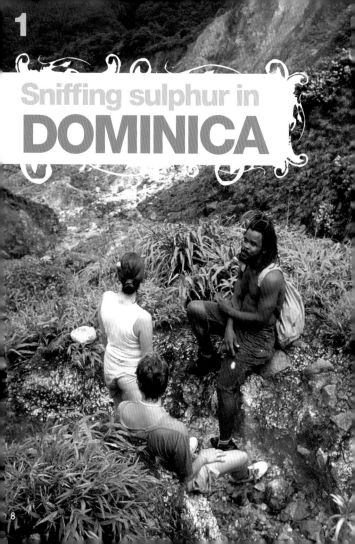

Sniffing sulphur in
DOMINICA

Reggae pumped from the speakers and lush foliage blurred through the windows of a Chinese-made minibus recklessly winding into the mountains at six in the morning. "Make sure you have enough water". "Dem shoes look like dey could melt". "Watch out your silver don't turn black". This sage advice came from Seacat, a Rastafarian guide to a small group of groggy but ambitious hikers about to trek to Dominica's famous Boiling Lake.

He grinned as he led them atop an elevated wooden water pipe that snaked through groves of the sweetest grapefruit on earth and clusters of familiar-looking office plants growing berserk in their native habitat. A trail shot uphill into the rainforest, a dense canopy of plants growing over plants that created a dark yet serene environment for the next two hours of climbing. Nearing the peak, the dense flora thinned into elfin trees and shrubs stunted by the powerful winds.

"Getting close", said the guide upon the first whiffs of sulphur, a reminder that an active volcano lurked nearby. The group clambered down into the caldera, appropriately named the Valley of Desolation for its scorched rock landscape. Mineral-encrusted fumaroles spewed boiling multi-hued gasses, sounding remarkably like a fleet of small jet engines. The group instinctively gathered together, some held hands, and everyone followed the guide's footsteps across this freakish landscape in which pterodactyls and dinosaurs would not appear out of place.

As the group inched toward a precipice overlooking a cauldron of bubbling muddy water nearly 70 metres across, they were enshrouded in steam and fog. Sulphurous gasses, reeking more powerfully than ever, had turned any silver jewellery to black. The guide climbed down to the edge of the water to boil a few eggs, proving a point.

On the return, hikers took a refreshing swim through a steep dark gorge to a small waterfall to rinse off the day's mud, soak weary muscles and postpone the white-knuckle ride out of the mountains and back to civilization.

need to know

To reach Dominica by air, you'll need to fly to another Caribbean island, such as Antigua or Barbados, to catch a connecting flight. Plenty of guides organize treks up to Boiling Lake; Seacat and his partner Roots also arrange excursions through the island. Contact them on ☎767/448-8954 or at ✉seacat55@hotmail.com.

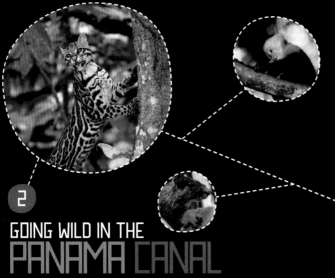

GOING WILD IN THE
PANAMA CANAL

Semi-hidden in the thick forest floor, a nervy agouti is on the forage for food. It stops to nibble on some spiny palm fruit, sandwiching each bite with anxious glances into the surrounding undergrowth. Less than five metres away from the unsuspecting rodent, and closing the gap with every stealthy stride, an ocelot moves in for the kill. The agouti enjoys its last few drips of sweet palm juice, and then...BANG. The cat gets the cream.

In the surrounding rainforest, several scientists from the Smithsonian Tropical Research Institute (STRI) monitor the moment, another important step in piecing together the relationship between ocelot and agouti, predator and prey. Further south, more scientists observe the family behaviour of white-faced capuchins; to the west, the evolution of Baird's tapirs is the focus. Welcome to Isla Barro Colorado, the most intensely studied tropical island on earth.

Sitting plum in the middle of man-made Lago Gatún, roughly halfway along the Panama Canal, Barro Colorado is a living laboratory, fifteen square kilometres of abundant biodiversity – more than a thousand plant species, nearly four hundred types of bird and over a hundred species of mammals – that the island owes to the canal itself. Flooding the area to facilitate its construction chased much of the wildlife in the surrounding

NEED TO KNOW

To visit Isla Barro Colorado, contact the **Smithsonian Institute** in Panama City (☎507/212 8026, ⊛www.stri.org). Guided tours (daily except Mon & Thurs; US$70, including lunch and boat transport to the island) last four to six hours; book well in advance as visitor numbers are strictly limited.

forests on to higher ground. The one-time plateau became an island, and the island became a modern-day Noah's Ark, the animals coming in two by two dozen to seek refuge in its dense rainforest covering.

The Smithsonian has run a research station on Barro Colorado for over eighty years, but only fairly recently have they opened their doors to tourists. It's been worth the wait, though – hiking through the rainforest in the company of expert guides, the jungle canopy abuzz with screeching howler monkeys and ablaze with red-billed toucans, is an incredible experience, even if the Hoffman's two-toed sloths, hanging lazily from the trees, seem distinctly underwhelmed by it all.

celebrating

FANTASY FEST in

KEY WEST

The saucy climax of Key West's calendar is a week-long party known as Fantasy Fest that caps the end of hurricane season in late October. That's when the old town is transformed into an outdoor costume bash, somewhat tenuously pegged to Halloween; really, it's a gay-heavy take on Mardi Gras, with campy themes (past ones have included "Freaks, Geeks and Goddesses" and "TV Jeebies") and flesh-flashing costumes.

The week is punctuated with offbeat events, like the pet costume contest where dogs and their owners dress the same and a sequin-spangled satire of a high school prom. On Saturday, the final parade slinks down the main drag, Duval Street, on a well-lubricated, booze-fueled route.

It's not surprising that such an irresistibly kitschy shindig should have emerged and endured in Key West. Save San Francisco, there's nowhere in America more synonymous with out and proud gay life than this final, isolated, all-but-an-island in the Florida Keys – the town's been a byword for homocentric hedonism since the sexually liberated 1970s. But in many ways, Key West's queer reputation is misleading. Sure, it's still a gay hotspot, as the thong-sporting go-go boys, who wield their crotches like weapons in bars along Duval Street, attest. But what drew the gay community here in the first place was the town's liberal, all-inclusiveness; be who you want to be, the locals say, whatever that is. It's all summed up by the town's official motto: One Human Family. Indeed, Key West welcomes everyone, from President Truman, who holed up in the so-called Little White House here in the 1940s to an oddball bar owner like "Buddy" today, who built his ramshackle café-cum-pub, B.O.'s Fish Wagon, out of piles of junk. After a few days, these kind of visual quirks seem quite standard – except maybe that pet costume contest.

need to know
Highway-1 is the single route linking Key West to the mainland, and a scenic, if long, drive; allow four hours each way to and from downtown Miami.
A better option is to book a one-way flight to Key West airport, which flies low enough to snap some impressive shots of the Keys from the air; and then amble back up by car along Highway-1, stopping as and when you wish. For more info on Fantasy Fest, check out ⊛www.fantasyfest.net.

4

A taste of Skye

The young waitress with soft skin and a lilting Hebridean accent handed over the menu and drew my attention to the day's specials chalked up on the blackboard above the mantelpiece.

I was in a window seat – though the windows of the rebuilt croft house were small, dating from days when glass was an inconceivable luxury and insulation was provided by thick stone walls and a thatch of reeds and bracken.My view included said reeds and bracken, growing thickly near the seaweed-tangled shore, a patch of steel-coloured water and green land beyond, rounded and speckled with sheep at first, then rising to broad-shouldered slopes and the more ragged shapes of higher hills beyond.

I hadn't expected the menu to compare to the scenery, but it did – and read like a mini guidebook. Crab fresh from Bracadale, langoustines caught in a Loch Dunvegan creel, scallops hand-dived near Sconser, wild venison from the Cuillin. There were woodland mushrooms, berries gathered from local gardens and doused in heather honey, creamy soft cheeses from Lochalsh and oatcakes made on a traditional flat skillet in the kitchen. Afterwards I sampled a tangy whisky distilled and aged in the salty coastal air at Talisker.

You come to a place as magical as Skye to discover the struggles of its history, the charm of its rugged shores, the drama of the jagged mountain tops and the seductive allure of the wonderfully green (but hardly fertile) landscape. You look out for a soaring golden eagle or a darting otter, tramp along mossy paths and scramble over tumbledown castle ruins. Skye fills the senses and tugs at your emotions, but it's a very pleasant realization that it can also fill your belly.

need to know

There are no scheduled air services to Skye. Trains run from Glasgow to Mallaig on the mainland, from where there's a ferry service to Skye, or from Inverness to Kyle of Lochalsh, linked by a road bridge to Skye. For a great meal, try the **Three Chimneys** (☎01470/511258, ⊛www.threechimneys.co.uk), **Kinloch Lodge** (☎01471/ 833333, ⊛www. kinloch-lodge.co.uk) or the **Stein Inn** pub (☎01470/592362, ⊛www.stein-inn.co.uk).

5

Living the quiet life in

LOFOTEN

Draped across the turbulent waters of the Norwegian Sea,
far above the Arctic Circle, Norway's Lofoten islands are,
by any standard, staggeringly beautiful.

In a tamed and heavily populated continent, the Lofoten are a rare wilderness outpost, an untramelled landscape of rearing mountains, deep fjords, squawking seabird colonies and long, surf-swept beaches. This was never a land for the faint-hearted, but, since Viking times, a few hundred islanders have always managed to hang on here, eking out a tough existence from the thin soils and cod-rich waters. Many emigrated – and those who stayed came to think they were unlucky: unlucky with the price of the fish on which they were dependent, unlucky to be so isolated and unlucky when the storms rolled in to lash their tiny villages. Then Norway found tourism and, though this started inauspiciously, when the first boatloads turned out to be English missionaries bent on saving souls, subsequent contacts proved more financially rewarding. Even better, the Norwegians found oil in the 1960s, lots and lots of oil, quite enough to extend the road network to the smallest village – the end of rural isolation at a stroke. All of the islands' villages have both benefited from this road-building bonanza and kept their erstwhile charm, from the remote Å i Lofoten in the south, through to the beguiling headland hamlet of Henningsvaer, extravagantly picturesque Nusfjord and solitary Stamsund.

Today, the Lofoten have their own relaxed pace and, for somewhere so far north, the weather can be exceptionally mild: you can spend summer days sunbathing on the rocks or hiking around the superb coastline. When it rains, as it does frequently, life focuses on the *rorbuer* (fishermen's huts), where freshly caught fish are cooked over wood-burning stoves, tales are told and time gently wasted. If that sounds contrived, in a sense it is – the way of life here is to some extent preserved for the tourists. But it's rare to find anyone who isn't less than enthralled by it all.

need to know
The Lofoten islands can be reached by car ferry, passenger boat and plane from Bodø, in northern Norway. There are flights to Bodø from many other Norwegian cities, including Oslo and Bergen.

17

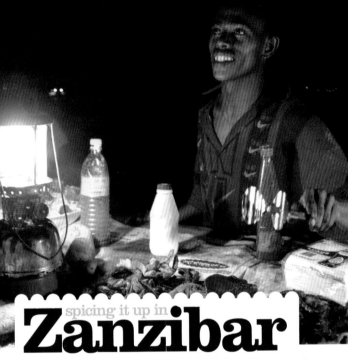

Zanzibar

Caressed by the warm waters of the Indian Ocean and cleansed by its monsoon, Zanzibar – East Africa's "Spice Islands" – feels worlds apart from the Tanzanian mainland just 40km away. A millennia of trade with lateen-rigged *dhows* (sailing vessels) introduced numerous peoples from faraway climes, all of whom contributed to the Swahili culture and language and also brought most of the ingredients that infuse one of Africa's most distinctive – and delicious – cuisines.

There's nutmeg and cloves from the Moluccas; cardamom, rice and pepper corns from India; aromatic Sri Lankan cinnamon; and sweet basil (and hookah pipes) from Persia. Portuguese caravels carried chili, vanilla and cassava from the Americas; Indonesians arrived with bananas, turmeric and coconuts; the Arabs introduced coriander and cumin; and Chinese fleets unloaded ginger, along with porcelain and silks for the wealthy.

A spice tour is de rigueur for visitors, but it's on the plate that Zanzibar's fragrant culinary marriage really shines. At nightfall in Stone Town's waterfront Forodhani Gardens, dozens of cooks set up trestle tables and charcoal stoves to prepare nightly banquets that would please any sultan, all in the flickering light of oil lanterns, and at bargain prices. Feast on octopus stewed in coconut sauce, along with fresh lobster, shrimp, prawns, king fish (diced and grilled), whole snappers and even shark. Cool your throat with tropical juices like coconut straight from the shell, tangy tamarind, mango, papaya, pineapple, banana and sugar cane – though you'll have to bear the banshee-like wails of an ancient iron press to sample that one.

Top off your meal with a tiny cup of Omani-style coffee laced with cardamom, and a glob of halua, a sticky, gooey confection made from wheat, pistachios, saffron and cardamom, and unbelievable amounts of sugar – the perfect sweet finale to a spicy feast.

need to know
Zanzibar is a twenty-minute flight from Dar es Salaam, and under three hours by ferry. You'll find plenty of beach hotels to suit all tastes and pockets, as well as luxuriously converted mansions in Stone Town.

6

"GET ON BAD"
IN TRINIDAD

Trinidadians are famed for their legendary party stamina, and nowhere is this dedication to good times more evident than in their annual Carnival, a huge, joyful, all-encompassing event that's the biggest festival in the Caribbean, and one which quite possibly delivers the most fun you'll ever have – period. And as Carnival here is all about participation – rather than watching from the sidelines à la Rio – anyone with a willingness to wine their waist and "get on bad" is welcome to sign up with a masquerade band, which gets you a costume and the chance to dance through the streets alongside tens of thousands of your fellow revellers.

Preceded by weeks of all-night outdoor fetes, as parties here are known, as well as competitions for the best steel bands and calypso and soca singers, the main event starts at 1am on Carnival Sunday with Jouvert. This anarchic and raunchy street party is pure, unadulterated bacchanalia, with generous coatings of mud, chocolate, oil or body paint – and libations of kick-ass local rum, of course – helping you lose all inhibitions and slip and slide through the streets 'til dawn in an anonymous mass of dirty, drunken, happy humanity. The air fills with music from steel bands, sound-system trucks or the traditional "rhythm section" band of percussionists. Once the sun is fully up, and a quick dip in the Caribbean has dispensed with the worst of the mud, the masquerade bands hit the streets, their followers dancing along in the wake of the pounding soca. This is a mere warm-up for the main parade the following day, however, when full costumes are worn and the streets are awash with colour. The music trucks are back in earnest and the city reverberates with music, becoming one giant street party until "las lap" and total exhaustion closes proceedings for another year.

need to know
The main parades in the capital, Port of Spain, take place on the Monday and Tuesday before Ash Wednesday. For info on accommodation and attractions in T&T, as well as Carnival, visit ®www.tntisland.com/mas.html. Also check ®www.carnaval.com, which has a list of mas band websites, where you can sign up online to play mas.

watching worlds collide
in SICILY

Dangling off Italy's toe at the centre of the Mediterranean, Sicily is a rich minestrone of ingredients – from smoking volcanoes to celeb-studded resorts, from screeching scooter boys and clamorous street markets to intense and emotional Catholic ceremonies.

The Med's largest island can boast vestiges of most of Europe's great civilizations, including majestic Greek temples and resplendent Norman mosaics. Sicily's capital, Palermo, is a distillation of the entire island, and in many ways the most idiosyncratic of Italy's regional capitals. In this full-on city set against a jagged mountain backdrop, you can experience the exotic, chaotic interface of Europe and North Africa.

The Arab flavour lingers on in such quarters as La Kalsa, in the dusty red cupolas of palm-encircled churches and in street markets that pullulate with energy, where you can pick up anything from a brace of still-flapping mullet to Rococo statuary.

Religious fervour manifests itself throughout the city in such raucous festivals as July's Santa Rosalia, while the churches offer extraordinary beauty, as in the glittering mosaics of Monreale or the swirling curlicues of the omnipresent Baroque. You can also experience the flamboyant flights of the Baroque imagination in the hundreds of elegantly crumbling palazzi, more often than not concealed behind stout walls, and sometimes – as in Bagheria – with dark, grotesque embellishments.

And of course, you can eat royally in this most sensuous of cities: tuck into *pasta con sarde*, the local speciality, in boisterous restaurants, and then gorge on cassata ice cream cake crammed with candied fruit.

If you're feeling stifled by the noise and confusion, escape is simple: sunbathe by day and jive by night at Palermo's seaside resort of Mondello; explore wooded Monte Pellegrino on a rented Vespa; or jump on a ferry to the offshore isle of Ustica, a marine reserve and diving paradise. Then, once revived, immerse yourself all over again.

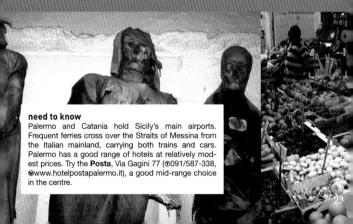

need to know
Palermo and Catania hold Sicily's main airports. Frequent ferries cross over the Straits of Messina from the Italian mainland, carrying both trains and cars. Palermo has a good range of hotels at relatively modest prices. Try the **Posta**, Via Gagini 77 (☎091/587-338, ⊛www.hotelpostapalermo.it), a good mid-range choice in the centre.

9 exploring
Malapascua
by banca

The view as you approach the tiny Philippine island of Mala-pascua by sea is one you'll never forget. It sits exquisitely in an emerald sea like some tremulous flying saucer, the palm trees of its jungled interior ringed by a halo of blindingly white sand. This is Bounty Beach, as idyllic as they come.

Difficult as it may be, summon the willpower to move from your shady hammock and explore beyond — you'll be further rewarded with picturesque inlets and coves where you can play ultimate castaway, with a few home comforts. The best way to explore is by hiring a *banca*, a native wooden pump-boat with spidery bamboo outriggers. A boatman and a couple of boatboys will come as part of the package, and they'll know exactly where to take you.

Start with Carnassa Beach, a secluded envelope of sugary sand banded by low hills on three sides with dwarf coconut palms for shade. On the fourth side is the sea, its gin-clear waters perfect for snorkelling among the reef fishes and the occasional inquisitive turtle.

From there chug north to Coconut Beach, a sunny crescent of sand backed by a coconut grove where you can drink fresh *buko* (coconut milk) straight from the tree and find a shady spot for a well-earned siesta.

For the ultimate sundowner, continue on to the unnamed curve of beach beyond the island's little lighthouse. From here you can gaze westward, chilled San Miguel beer in hand, as the tropical sun slips lazily below the horizon. South of here is Logon Beach, where the locals will rustle up some grilled fish for your dinner. Wash it down with a grog or two of local rum, and spend the night in simple huts made from native grass.

Next morning, set out at dawn for the two-hour *banca* trip to Calangaman, a stunningly gorgeous dumbbell-shaped slither of sand seemingly in the middle of nowhere. The islet's only residents are a handful of fisherfolk who can help you prepare an unparalleled "catch of the day" beach barbecue. Just don't forget the cold drinks.

need to know
Daily flights from Manila and a number of Southeast Asian cities head to Mactan Cebu International Airport, the closest airport to Malapascua. From there you can rent a car and driver or take a bus to Maya, the northernmost town on mainland Cebu. Regular ferries run from Maya to Malapascua.

10

A TOAST TO VIKINGS IN BORNHOLM

Thick smoke wafts over the slender herring, rows of them, as their silvery scales warm into a golden-red. The flushed Dane, in coveralls and clogs, prods the alderwood embers with a long pole swathed in rags at one end. Inside the smokehouse, it's damp and dark, not much larger than a garden shed – and just as basic. Then again, so is the herring preparation – here on the wave-lashed Danish island of Bornholm, this tradition of fish meeting fire owes a debt to the island's first Viking inhabitants, who pulled up in their longboats a millennium ago.

The petite kingdom has come a long way. These days, its bright ferries filled with sun-seekers that pull up – to an island that embodies Denmark's penchant for all things *hygellig*, or "cozy and warm": brick-tiled roofs top custard-yellow half-timber houses, lace curtains frame dollhouse windows and in the quiet harbour, fishing boats bob to the squawks of gulls circling lazily above. Off in the distance, slender smokehouse chimneys punctuate the low-rise landscape, snorting smoke into the bright North sky.

Emerge from the tangled Almindingen forest and you'll come across places like Gudhjem, or "God's Home" – just what you might expect if the Man Upstairs were to design his perfect village, especially when the last gasp of sunlight strikes the cobblestone streets. The island's twelfth-century round churches (*rundkirke*) – whitewashed fortresses capped with ink-black conical roofs – lend a stylized, medieval splendour to the otherwise tidy pastureland.

While the Vikings' table "manners" likely raised a few eyebrows – they didn't use plates or utensils but for the knives they pulled from their sheaths – today, or so the saying goes, the only time you'll see a Dane with a knife in hand is when he has a fork in the other. Still, as you feast on "Sun over Gudjhem" (smoked herring on dark bread, topped with a quivering egg and raw onions), washed down with glass after glass of chilled Tuborg, you may feel some distant connection to those helmet-wearing voyagers. They did know a thing or two about having a good time.

need to know
Ferries travel daily to Bornholm from Ystad, in southern Sweden (1hr 15min), and from Køge, south of Copenhagen (overnight). You can also catch the daily Bornholmer bus from Copenhagen (2hr 45min). For further information, check out ⊛www.bornholmferries.com.

11
dawn over
Kelimutu
in Flores

There's something magnificently untamed about Flores. As with almost every island in this part of the Indonesian archipelago, Flores is fringed by picture-postcard beaches of golden sand. But to appreciate its unique charms you have to turn away from the sea and instead face the island's interior. Despite its relatively small size (a mere 370km long and, in places, as narrow as 12km), only the much larger Indonesian landmasses of Java and Sumatra can boast more volcanoes than this slender sliver of lush land.

Unsurprisingly, perhaps, Flores' jagged volcanic spine has played a major part in the island's development. The precipitous topography contributes to the island's torrential wet seasons, which in turn provides a tropical countenance – not for nothing did the Portuguese name this island "Cabo das Flores", the Cape of Flowers. The rugged peaks have also long separated the various tribes on the island – an enforced segregation that has ensured an inordinate number of different languages and dialects, as well as many distinct cultures.

The highest volcanic peak on the island is the towering 2382m Gunung Ranaka, while its most volatile is the grumbling, hot-headed Ebulobo on Flores' south coast. But the prince among them is Kelimutu.

Though just 1620m high, the volcano has become something of a pilgrimage site. Waking at around 4am, trekkers pile onto the back of an open-sided truck for the thirty-minute ride up the volcano's slopes, from where a short scramble to Kelimutu's barren summit reveals the mountain's unique attraction: three small craters, each filled with lakes of startlingly different colours, ranging from vibrant turquoise to a deep, reddish brown. With the wisps of morning mist lingering above the water's surface and the rising sun bouncing off the waters creating an ever-changing play of light and colour, dawn over Kelimutu is Indonesia at its most beguiling.

need to know
Regular flights from Bali serve Flores' three main airports, Labuanbajo, Maumere and Ende. The early-morning truck to Kelimutu travels from Moni to the foot of the final climb to the summit; it returns to Moni at 7am sharp.

The seasons of the Scillies

It's no exaggeration to say that, from London, you can get to the Caribbean quicker than the Scillies. But then, that's part of the appeal. You take the night train down to Penzance and then hop on the Scillonian ferry (not for the queasy) or the helicopter out over the Atlantic.

What awaits depends very much on the weather. The Scillies out of season are bracing, as wind and rain batters these low island outcrops. Wrapped in waterproofs, you can still have fun, squelching over wet bracken and spongey turf to odd outcrops of ruined castles, or picking through the profusion of shells on the beaches. And if you can afford a room at the **Hell Bay Hotel** on Bryher, then you can top off the day with a first-rate meal while gazing at original sculptures and paintings by Barbara Hepworth and Ivor Hitchens.

In summer, when the sun's out, it's a very different scene, and you can swim, go boating, even learn to scuba dive – the water is crystal clear and there are numerous wrecks. It can be a cheap holiday, too, since for only

a few pounds you can pitch a tent at the Bryher campsite and enjoy one of the loveliest views in Europe. From there, wander down to the **Fraggle Rock** pub for a pint of Timothy Taylors and a crab sandwich, and then, if the tide's right, wade across to the neighbouring island of Tresco, and explore the subtropical Abbey Gardens.

I keep stressing Bryher, as that's my island. Scilly-fans are fiercely loyal. But you can have almost as good a time on St Martin's, which has arguably the best beaches, and a brilliantly sited pub, **The Turk's Head**, or on the diminutive St Agnes (population 72), with its wind-sculpted granite. St Mary's doubtless has devotees, too, but with the islands' main town, regular roads and cars, it lacks essential island romance. For that, you'll find me on Bryher.

need to know

From June to September, you need to book well ahead for accommodation – and for the helicopter. Hotel and B&B rooms are scarce, as are rental cottages; campsites do generally have space, though you'll need bring along camping gear. **The Hell Bay Hotel** on Bryher (☎01720/422947, ⊜www.hellbay.co.uk) offers rooms (meals included) for £150–250.

13
Soaking in **Iceland's** hot springs

Most people visit Iceland in summer, when once or twice a week it actually stops raining and the sun shines in a way that makes you think, briefly, about taking off your sweater. The hills show off their green, yellow and red gravel faces to best effect, and you can even get around easily without a snow-plough. But if you really want to see what makes this odd country tick, consider a winter visit. True, you'll find many places cut off from the outside world until Easter, people drinking themselves into oblivion to make those endless nocturnal stretches race by (though they do they same thing in summer, filled with joy at the endless daylight) and tourist information booths boarded up until the thaw. On the other hand, you can do some things in winter which you'll never forget.

Up in the northeast, Lake Myvatn is surrounded by craters, boiling mud pools and other evidence of Iceland's unstable tectonics. None of which is better than the crevasses up near the northeast shore, flooded by thermal springs welling up out of the earth. Too hot for summer bathing, in winter the water temperature drops to just within human tolerance, and they're best visited in a blizzard, when you'll need to be well rugged up against the bitter, driving wind and swirling snow. Clamber up the steep slope and look down over the edge: rising steam from the narrow, flooded fissure five metres below has built up a thick ice coating, so it's out with the ice axes to cut footholds for the climb down to a narrow ledge, where you undress in the cold and, shivering, ease yourself into the pale blue water. And then... heaven!

You tread water and look up into the falling snow and weird half-light, your damp hair nearly frozen but your body flooded with heat. Five minutes later, you can't take any more and clamber out, face red as a lobster and your body feeling so hot you're surprised that the overhanging ice sheets haven't started to melt.

need to know

Myvatn is about six hours by road from Iceland's capital, Reykjavík, and about two hours from the nearest town, Akureyri. Buses run from either in summer (approximately June–September), but you'll have to hire a car during the rest of the year.

14

HIKING THE ANCIENT FORESTS OF LA GOMERA

Though an easy ferry ride from Tenerife, La Gomera, the smallest of Spain's Canary Islands, is one of Europe's most remote corners. Indeed, this is where a number of 1960s American draft dodgers sought refuge, and it remains a perfect place to get away from it all. In its centre unfold the ancient forests of the Parque Nacional De Garajonay – a tangled mass of moss-cloaked laurel trees thriving among swirling mists to produce an eerie landscape straight out of a Tolkien novel.

The park is best explored along the rough paths that twist between the many labyrinthine root systems. Embark on its finest trek, a 9km trek that's manageable in about four hours if you use a bus or taxi to access the start (a road intersection called Pajarito) and finish points. The lush route takes in the island's central peak, Garajonay; from its summit, you can enjoy immense views looking out over dense tree canopy to neighbouring islands, including Tenerife's towering volcano Mount Teide – at 3718m, Spain's highest point. From Garajonay's peak, follow a crystal-clear stream through thick, dark forest to the cultivated terraces around the hamlet of El Cedro.

Here you can camp, get a basic room or even rent a no-frills cottage – but be sure to at least pause at its rustic bar. Settle onto a wooden bench and try some thick watercress soup sprinkled with the traditional bread-substitute *gofio*, a flour made from roasted grains.

Beyond El Cedro the valley opens up to the craggy and precipitous landscape that surrounds the town of Hermigua. The terrain around here is so difficult that for centuries a whistling language thrived as a means of communication, but today catching a bus back to your base – via dizzyingly steep hillside roads – is an easy matter.

need to know
La Gomera lies 28km from Tenerife and is linked by at least ten ferries per day to its capital San Sebastián (45min; around €20 one-way). Easily manageable on foot, San Sebastián has a good selection of basic pensions, standard hotels and a luxury hotel.

DWARFED BY HONG KONG'S SKYSCRAPERS

If there's anywhere in the world that can cut you down to size it's Central district, on the north shore of Hong Kong Island. Walking around here you're humbled by just about everything – the crowds, the incomprehensible language which eveyone else seems to have mastered, the heat, the superbly enticing smells of Chinese cooking and the dazzling beauty of Hong Kong harbour, dotted with ferries and tankers. More than anything, though, it feels like being at the bottom of a well: this is Asia's financial hub, and Hong Kong is a place where size really does matter, so building a one-story, local branch-style financial institution is simply not an option. Skyscrapers rear up all around, so closely packed that their lower stories have been linked by a network of pedestrian walkways.

The tallest building here, Tower Two of the International Finance Centre, literally touches the clouds at 420m high. Nonetheless, local boys Bank of China (BOC) and the Hong Kong and Shanghai Banking Corporation (HSBC) have been slugging it out for decades over who can build the most auspicious building. And here it's not just size alone that counts, but the Chinese belief in feng shui, or how the different parts of a landscape influence each other. HSBC is built off the ground – you can walk right underneath it and look up at the offices surrounding the hollow central well – because otherwise it would have blocked the flow of "good luck" between Government House (further uphill behind it) and the waterfront. Built in the 1980s, it's gargantuan next to the BOC of the time, but the Chinese picked up their act and in the 1990s built a new bank headquarters which is not only so high that the head of the BOC can look down on his competitor, but also puts feng shui to some nasty use – the new BOC tower points skywards like a knife, stabbing the heavens and pulling out its good luck.

need to know

Central is on the north shore of Hong Kong Island, near the tourist hubs of Kowloon (directly north across the harbour) and Causeway Bay (about 2km east). Central is easily reached from anywhere in Hong Kong via the MTR (subway), though it's also a scenic boat ride from Kowloon on the Star Ferry.

need to know
Virgin Atlantic and Cubana offer direct flights to Cuba from the UK. There are no scheduled flights from the US to Cuba but charter flights travel from Miami, New York and Los Angeles. **La Guarida**, at Concordia 418, e/ Gervasio y Escobar, Centro Habana (☎7/863 7351 and 866 9047), is open daily, noon to midnight.

CLASSIC CARS & CUISINE IN CUBA

16

There's no more striking reminder of the relationship, once a tempestuous love affair but long since turned rancorous, between Cuba and the US than at the Parque de la Fraternidad on the edge of Habana Vieja. The Capitolio – the Cuban replica of Washington's Capitol Building – dwarfs the surrounding nineteenth- and early twentieth-century balconied buildings, many faded and crumbling under the weight of the sheer number of families living inside them.

On the park roads, you'll see lines of Buicks and Chevrolets, Oldsmobiles and Chryslers, some apparently still in their heyday fifty years on, most looking more their age. They're all taxis – and their numerous ranks are one of the more prominent examples of the market flexing its muscle in this Communist stronghold.

Your destination is **La Guarida** – and the journey is a short one. Once you've negotiated a price, the taxi driver negotiates the streets, clattering over pot holes on a ride that takes you from colonial Habana Vieja to neo-colonial Centro Habana, a dusty mass of apartment buildings enclosing shadowy streets, where kids play bottle-cap baseball amid the chatter of doorstep politics.

La Guarida is one of the capital's best **paladares**, the family restaurants allowed to compete with the state run monopoly. As the taxi pulls up, you may first think there has been some mistake. All that's visible is a dingy hallway and a dirty marble staircase that winds away out of sight. Even after scaling the first flight of steps, which lead to a semi-derelict first floor, you may still be in doubt. Then, on the third floor, a door opens onto a small cloakroom that gives way to what was clearly once someone's home. The owners have gotten away with providing more than twelve chairs for their diners, usually the legal limit for a private eatery, which is just as well, as the popularity far outstrips the capacity. And for good reason: tucking into some of the finest Habana cuisine at the top of a crowded decrepit apartment building offers an unforgettable glimpse into today's Cuba.

trekking the
Kalalau Trail
on Kauai

Kauai is the Hawaii you dream about. Spectacular South Seas scenery, white-sand beaches, pounding surf, laid-back island life – it's all here. While the other Hawaiian islands all have the above to varying degrees, none has quite the breathtaking beauty nor sheer variety of beguiling landscapes like Kauai. And none has a shoreline as magnificent as the Na Pali coast (which means "the cliffs" in Hawaiian), where gorgeous lush valleys are separated by staggering knife-edge ridges of rock, towering over 900 metres tall and clad in glowing green vegetation that makes them resemble vast pleated velvet curtains.

The one road that circles Kauai peters out on the North Shore. Shrinking ever narrower to cross a series of one-lane wooden bridges, it finally gives up where lovely Ke'e Beach nestles at the foot of a mighty mountain. Swim out a short distance here, and you'll glimpse the mysterious, shadowy cliffs in the distance, dropping into the ocean.

Now deserted, the remote valleys beyond the end of the road once supported large Hawaiian populations, who thought nothing of riding the fearsome waves in their canoes. The only way to reach them these days is on foot. One of the world's great hikes, the 17km Kalalau Trail is a long, muddy scramble, on which one moment you're perched high above the Pacific on an exposed ledge, and the next you're wading a fast-flowing mountain stream. Your reward at the far end – an irresistible campsite on a long golden beach, where you can breathe the purest air on Earth, explore the tumbling waterfalls of Kalalau Valley and then gaze at the star-filled sky into the night.

need to know
Direct flights connect Lihue on Kauai with Honolulu, San Francisco and Los Angeles. Of the hotels on the North Shore, the pick of the bunch is the ravishing oceanfront **Hanalei Colony Resort** (☎808/826-6235 or 1-800/628-3004, ⊛www.hcr.com).

Relaxing in tropical
TAKETOMI-JIMA

18

On tiny Taketomi-jima the traditional bungalow homes are ringed by rocky walls draped with hibiscus and bougainvillea. From the low-slung terracotta-tiled roofs glare *shiisa*, ferocious, bug-eyed lion figures. The only traffic is cyclists on rickety bikes negotiating the sandy lanes, and buffalo-drawn carts, hauling visitors to the beaches in search of minuscule star-shaped shells.

This is Japan – but not as you might know it. Taketomi-jima is one of the hundred-plus sub-tropical islands of Okinawa that trail, like scattered grains of rice, some 700km across the South China Sea. Cultural influences from China, Southeast Asia and the US, who occupied Okinawa until 1972 following WWII, have all seeped into the local way of life, providing a fascinating counterpoint to the conformity and fast-paced modernity of mainland Japan.

Fringed by soft, golden beaches, the islands are popular with Japanese looking for some R & R. Taketomi-jima especially is often besieged with daytrippers, as it's just a short ferry ride from the mountainous Ishigaki-jima, the main island of the Okinawan sub-collection known as the Yaeyamas. The trick is to stay on after the masses have left. Take up residence in one of the many family-run *minshuku* – small guesthouses with tatami mat floors, futons and rice-paper *shoji* screens. After a refreshing bath, slip into the provided *yukata* (cotton robe) and dig into a tasty dinner of local delicacies, including tender pork and fresh seafood, and then wander down to the beach to watch the glorious sunset.

On returning to the *minshuku*, it's not unusual for a bottle or two of Okinawa's pungent rice liquor *awamori* to appear. Locals strum on a *sanshin* (three-stringed lute) and lead guests in a gentle sing-along of Okinawan folk favourites. As the *awamori* takes effect, don't be surprised if you also learn a few local dance moves.

need to know

Taketomi-jima is reached by ferry from Ishikagi on Ishikagi-jima, which has direct flights to Tokyo and Osaka. For more info, contact the local tourism association on ☏0980/82-5445, or go to the Japan National Tourist Organisation's homepage (🌐www.jnto.go.jp).

SEEING LEGENDS
COME ALIVE IN CRETE

Crete has never been just another Greek island. True, it has everything you could ask for – crystalline waters, sun-washed beaches, isolated villages and hospitable people. But it is also much more than that.

On Crete, legends come alive. Here, Homer's stories of King Minos emerged from myth and were proven to be fact; Zeus was born here (and died here too, some Cretans claim); and four thousand years ago the first civilization in Europe flourished on this island, while the rest of Greece was just learning to speak Greek.

Crete's size and position – as close to Africa as to Athens, a meeting point of east and west – has meant a history of exceptional richness, a history that has moulded the islanders and forged a spirit of indomitable resistance. There are mountain villages where you can see the traces of all those 4000 years of continuous human occupation, and the many invaders and locals who have left their mark.

If you want, you can swim and lie by the beach and know nothing of any of this. But even so, it's hard to ignore Crete's bold physical presence – its great mountain ranges, snow-capped well into the summer, dominate the inland landscape. Cutting through them are impressive gorges, most famously at Samariá, that offer fabulous opportunities for walkers. At the mountains' feet lie the great Minoan palaces, but also lesser-known relics: Roman cities; frescoed Byzantine churches; Venetian and Turkish fortresses.

Even the food is redolent of tradition, the wines, olive oil and simple fare tasting of the Mediterranean sun and Cretan soil – linking you closer to this fertile land of legends.

need to know
As the southernmost Greek island, Crete has long summers, and a tourist season that runs from April to October. There are direct flights to the two main airports at Iráklion and Haniá from the UK and many northern European cities. Otherwise fly to Athens, from where it's a half-hour flight or comfortable overnight ferry ride.

20
HIGH IN HAPUTALE

Sri Lanka has many unexpected sights, but few as surreal as early morning in Haputale. As dawn breaks, the mists that blanket the town for much of the year slowly dissipate, revealing the huddled shapes of dark-skinned Tamils, insulated against the cold in woolly hats and padded jackets, hawking great bundles of English vegetables – radishes, swedes, cabbages and marrows – while the workaday Sri Lankan town slowly comes to life in the background, with its hooting buses and cluttered bazaars. As the mists clear and the sun rises, the tangled ridges of the island's hill country come slowly into view to the north, while to the south the land falls dramatically away to the lowlands below, with the far-off view of the coast and its sweltering Indian Ocean beaches faintly visible in the distance. As an image of Sri Lanka's unexpected juxtapositions, Haputale has few peers, and to stand shivering on a hilltop within a few degrees of the equator, watching a scene reminiscent of an English market town crazily displaced in time and space, is to understand something of the cultural and physical contradictions of this fascinatingly diverse island.

The contradictions continue in the countryside beyond Haputale, as the road twists and turns up into the sprawling British-era plantations of the Dambatenne Tea Estate, whose antiquated factory is filled with the ingenious Victorian mechanical contraptions which are still used to process the leaves brought in from the surrounding estates. For the British visitor particularly, there is always the faint, strange nostalgia of seeing the legacy of one's great-great-grandparents preserved in a distant and exotic tropical island. But there is also the subversive awareness that the hillsides of Haputale, once colonized by British, have now reached out and quietly conquered distant parts of the world in their turn, filling the tea-bags and chai shops of countries as varied as England, Iran and India with a taste which is purely and uniquely Sri Lankan.

need to know
Haputale can be reached by direct
trains from both of Sri Lanka's two
principal cities, Colombo (9hr) and
Kandy (5hr 30min). Accommodation in the town is limited to a
handful of low-key but comfortable
guesthouses, the best of which is
the excellent **Amarasinghe Guest
House** (☎075/226-8175).

21
Whale-watching in the
Dominican Republic

There's no Caribbean island experience that tops sitting on a seaside verandah in the sleepy town of Samaná, on the Dominican Republic's northeast coast, and sipping a *cuba libre servicio* – two Cokes, a bottle of aged rum and a bucket of ice – as you watch a series of massive humpback whales dive just offshore. Thousands of whales – the entire Atlantic population – flock to the Samaná waters each winter to breed and give birth. And no matter how long you relax there looking out at the swaying palms backed by a long strand of bone-white sand, you never cease to be surprised as one whale after another sidles up the coastline and emerges from the tepid depths of the Samaná Bay before coming back down with a crowd-pleasing crash.

Sleepy Samaná is refreshingly free from package tourists. The modest expat community is almost all French, and they've set up a series of laid-back outdoor eateries along the main road. A lot of the native Dominicans are from the United States originally – free blacks from the time of slavery who moved here in the early nineteenth century when it was a part of Haiti, the world's first black republic.

Samaná also once held an allure for the Emperor Napoleon, who envisioned making its natural harbour the capital of his New World empire. While he never carried out his grand plans, look out over Samaná today and you can still imagine the great Napoleonic city destined to remain an Emperor's dream: a flotilla of sailboats stands at attention behind the palm-ridged island chain, and in place of the impenetrable French fortress that was to jut atop the western promontory is a small, whitewashed hotel.

For now though, Samaná remains passed over, which means you can have its natural beauty, tree-lined streets and, above all, its spectacular whale population, pretty much all to yourself.

need to know
Fly into Puerto Plata's international airport, from where it's a four-hour bus ride to Samaná village. Check with **Victoria Marine** (☎809/538-2494) for a whale-watching boat tour.

ROAMING THROUGH RAINY PENANG

It's tropically hot and raining in Penang's old town centre, Georgetown, the water hitting the road so hard that it forms a haze at about knee height.

People are jumping the flooded gutters and running for cover under the collonaded shopfronts but there's not much room thanks to the squadrons of motorbikes parked here. Under this heavy downpour, you couldn't imagine a better evocation of colonial outpost gone to seed: wet, narrow lanes link the faded, mildewed buildings, their tops sprouting moss and small bits of vegetation.

At first, Georgetown gives little away about the people who actually live here; you can see the same century-old buildings all across southeast Asia from here to the Philippines. But as you wander about you begin to notice the distinct shapes of the temples, the unique scripts and the different market and restaurant smells that mark out each community of the Chinese, Hindu, Burmese and Thai settlers. Weirdly, the British – in control here for so many years – and the indigenous Malay seem barely represented; just the seafront skeleton of Fort Cornwallis and plush Eastern & Oriental hotel, and a night market where Malay youths parade their motorcycles and run stalls selling *Nonya* (Malay-Chinese) dishes.

About the only thing tying the communities together on this betel nut-shaped island (after which it's named) is the hot, sticky weather, and the way that everyone here seems to be under constant threat from the jungle that lurks in the background, just waiting to reclaim the city. Ride the funicular railway to the top of Penang Hill and you can see the reality of this: Penang's modern sprawl of towerblocks and expressways is laid out beneath you but up here there are huge trees, orchids, groves of bamboo and palms, greenery and red earth – not the wilds exactly, but only one step away from it. Down at the base of the hill, a walk around the Botanic Gardens only reinforces the feeling; here there are even monkeys, snakes and flying lizards. You can't help feeling that despite the development, Penang never really got anywhere, or at least, never fulfilled the first European colonists' expectations – though maybe it's just the heat and the heavy, waterlogged air.

need to know

Penang lies just off Malaysia's west coast, connected to the nearest mainland centre of Butterworth by a bridge; there are also domestic and international flights to the island. For more information, check out ⓦwww.tourismpenang.gov.my.

23

First impressions of Fogo, rising in a cone 2800 metres above the ocean, are of its forbidding mass, the steep dark slopes looming above the clouds. The small plane is buffeted on Atlantic winds as you dip onto the runway, perched high on dunnish cliffs above a thin strip of black beach and ultramarine sea.

Fogo ("Fire" in Portuguese) is the most captivating of the Cape Verde islands, with its dramatic caldera and peak, its fearsomely robust red wine, the island's traditional music and the hospitable Fogo islanders themselves.

Up in the weird moonscape of the caldera, after a slow shared taxi ride from the picturesque capital, São Filipe, you'll be dropped on the rocky plain at a straggle of houses made of grey volcanic tufa. One of a handful of charmingly rakish-looking guides will greet you, host you in his own home and, early next morning, take you up the steep path to the summit, Pico de Fogo, with its extraordinary panorama. The descent is better than any theme park ride as both

Peaking on Fogo
in Cape Verde

guide and guided leap and tumble through the scree in an entertaining, inelegant freestyle.

Shaking out the dust and volcanic gravel at the bottom, you'll fetch up at the local social club where an impromptu band – guitar, violin, keyboard, scraper and *cavaquinho* (four-stringed mini-guitar) – emboldened by the consumption of much *vinho* and local hooch or grogue, often strikes up around the bar. Wind down your evening listening to Fogo's beautiful *mornas* – mournful laments of loss and longing that hark back to the archipelago's whaling past and to families far away.

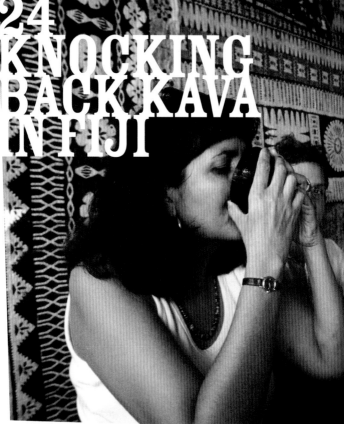

24 KNOCKING BACK KAVA IN FIJI

need to know

Nadi, on Viti Levu, is a major South Pacific hub, with numerous international flights, while the capital, Suva, on the southeast coast, is served by a smaller airport. **Home Stay Suva** (265 Prince's Rd, ☏679/337 0395) is the best deal in the capital, a colonial home with magnificent views.

FIRST, YOU'LL BE INVITED TO JOIN THE GROUP, LANGUIDLY ASSEMBLED AROUND A LARGE WOODEN BOWL.

Then, a grinning elder will pass you a coconut shell, saying "tovolea mada" – "try please". You take a look – the muddy pool in the shell looks like dirty dishwater, but what the hell, you sip anyway. And then the taste hits you, a sort of medicinal tonic tinged with pepper. Resist the urge to spit it out and you'll gain the respect of your hosts. Keep drinking, and you'll start to get numb lips, feel mildly intoxicated and if you're lucky, end up as tranquil as your new friends.

It's hard to forget the first time you try kava. Known as *yaqona* in Fiji, it's made from fresh kava root, and has been imbibed for hundreds of years, remaining a potent link to the island's past. For the locals, sharing kava is long-time social ritual, and it's common for families and friends to drink together at weekends or special occasions – getting invited to join them is one of the island's highlights.

Fiji's past also once included cannibalism, but visitors have long been off the menu, and you can safely take a trip to one of the traditional villages, such as Navala on the main island of Viti Levu, where you'll see houses, or *bure*, made of bamboo and leaves. The island's rich cultural heritage includes a blend of Polynesian and Melanesian, with nearly half the population of Indian origin. Suva, the capital, is filled with curry houses and even Hindu temples. Fiji became a British colony in 1874 (gaining independence in 1970), and you'll come across numerous evocative colonial remains straight out of a Graham Greene novel, along with magical church choirs that are a harmonious blend of Victorian melodies and tribal soul. Still, if you want to get to the heart of Fiji, drinking kava is a good place to start.

GREETINGS FROM

NANTUCKET ISLAND

MASS

25
gazing at a
NANTUCKET SUNSET

1850

You spent the morning beside a peppermint-striped lighthouse, and then meandered past shingled, rose-covered cottages. For lunch, you feasted on freshly steamed lobster with sweet corn on the cob. But in truth, you only really feel like you've arrived in Nantucket once you've sauntered along one of its long stretches of sugary sand, with the moorlands and breezy dunes to your back, and to the west, a gorgeous, swirly-pink sun slipping into a glowing sea. Sunset in Nantucket is a near-spiritual affair, and here, surrounded by sky and sea, you remember again why you have journeyed so far to a place that's so small.

Difficult to access – it's a two-hour, no-land-in-sight boat ride from the mainland – and primarily a summertime destination, the tiny island is imbued with a catch-it-while-you-can ambience. It's as if the sun and surf have filtered out the details of life on the mainland, shaping the island into a quiet, sandy idyll. There's not really much "to do" in Nantucket, but that's where its charm lies – content yourself with ice cream cones and seashells, salty sea air and bike rides to the beach.

As the evening fog rolls in – the island is famous for it, even owing its nickname, "the Grey Lady", to the mist – beach-weary folk head for Nantucket Town, packing into the drugstore for creamy milkshakes from its time-worn soda fountain where the stools still spin.

Nantucket wasn't always a land of leisure. During the eighteenth and nineteenth centuries, it served as a hub for the whaling trade, a dangerous industry marked by meager pay and grueling years at sea. Evidence of this hard-working heritage is well-preserved in the pretty sea captains' homes and cobblestones that line Main Street, restorative efforts largely funded by the island's well-heeled summer residents.

More than anything, windswept Nantucket is a wistful sort of place. As you board the ferry to go home, you'll long not only for simpler days, but also for the chance to come back again.

need to know

The Steamship Authority (☎508/693-9130, ⊛www.steamshipauthority.com) and **Hy-Line Cruises** (☎1-800/492-8082, ⊛www.hy-linecruises.com) run ferries to Nantucket from Hyannis, Massachusetts.

25

Ultimate
experiences
Islands
miscellany

Largest & smallest

The world's largest island is Greenland (Australia is considered the smallest continent, rather than the largest island), while Bishop Rock (a mere 46 x 16 metres) on the western edge of the Isles of Scilly (UK), is the smallest. The most populous is Java (Indonesia) with 124 million people; at the opposite of the spectrum is Pitcairn in the South Pacific, with just 67 residents.

The big five	Size (square kilometres)
Greenland	822,706
New Guinea	785,753
Borneo	748,168
Madagascar	587,713
Baffin Island (Canada)	507,451

 # Guano

For hundreds of years Pacific islands such as Nauru and Peru's Chincha islands have protduced heaps of guano (the collected droppings of seabirds and bats), a much sought-after fertilizer with high levels of phosphorus and nitrogen. It was deemed so valuable that the US Congress passed the Guano Islands Act in 1856, which made it mandatory for any US citizen discovering an island or rock containing guano to hand it over to the US government (provided it didn't belong to any other nation).

 # Faux isles

Of numerous artificial island projects around the world, Palm Islands, off the coast of Dubai, is the largest and most ambitious, involving three manmade islands shaped like palm trees and plastered with luxury hotels and holiday homes. Palm Deira will be the largest of the three, 14km long and 8.5km wide, bigger than Manhattan.

Divided islands

The partition of islands into separate political entities has often led to conflict. New Guinea is split between Papua New Guinea and Western New Guinea, administered by Indonesia, though many locals call it West Papua and have been struggling for independence. Ireland was partitioned in 1922 into The Republic of Ireland and Ulster, or Northern Ireland, which remains part of the UK – the IRA conducted a terrorist war against the latter until 1997. East Timor only became independent from Indonesia, which still controls West Timor, in 2002. Cyprus has been bitterly divided into Turkish and Greek zones since 1974.

"I am a rock, I am an island… And a rock feels no pain, and an island never cries."

Paul Simon, singer and songwriter

 # Private getaways

Island	Buyer
Blackadore Cay, Belize (104 acres)	Leonardo DiCaprio in 2005 for unconfirmed US$1.75–2.4m
Leaf Cay, Bahamas (25 acres)	Nicolas Cage in 2006 for a reputed US$3m
Little Hall's Pond Cay, Bahamas (45 acres)	Johnny Depp in 2004 for US$3.2m
Mago Island, Fiji (22sq km)	Mel Gibson in 2005 for US$15m
Necker Island, British Virgin Islands (74 acres)	Richard Branson in 1978 for US$303,000 (currently valued at around US$106m).

6 Literature

Islands have long featured in literature since Homer's *Odyssey* charted the eponymous hero's journey through the Aegean Sea. Shakespeare set his most philosophical play, *The Tempest*, on an island, and ever since John Donne told us that, "no man is an island, entire of itself", poets have been using them as symbols of isolation.

Born on St Lucia, Nobel-prize winner Derek Walcott is best known for his epic poem *Omeros*, which charts a fictional journey through the Caribbean. Somerset Maugham's fictional homage to Gauguin, *The Moon & Sixpence*, was inspired after his visit to Tahiti. One of the most popular children's books, *Anne of Green Gables* by Lucy Maud Montgomery, was set on Canada's Prince Edward Island.

Other books that immortalize islands include *Zorba the Greek* by Nikos Kazantzakis, which is set on rustic Crete; Michael Ondaatje's *Running in the Family*, a vivid depiction of life in Sri Lanka; the works by Halldór Laxness, who won the Nobel Prize for his powerful satires on modern Iceland; and The Bone People, a novel exploring Maori myths by Keri Hulme, one of New Zealand's most celebrated authors.

▶▶ Five island classics

Robinson Crusoe by Daniel Defoe. The original desert island novel.

Lord of the Flies by William Golding. Masterful, disturbing depiction of moral breakdown among a group of schoolboys stranded on a tropical island.

The Beach by Alex Garland. *Lord of the Flies* meets the gap-year generation.

Island by Aldous Huxley. Classic story of a utopian island society, threatened by an envious outside world.

Treasure Island by Robert Louis Stevenson. One of the world's best loved stories, with pirates, parrots and buried chests of gold.

"Fifteen men on the dead man's chest –
Yo-ho-ho, and a bottle of rum!"

Robert Louis Stevenson, Treasure Island

Party islands

Ibiza, one of Spain's Balearic islands, has been party central since the 1960s, assuming iconic status in the 1990s with the advent of super clubs such as Pacha, Space, Amnesia and Privilege, currently the world's biggest. Greek islands have also garnered a reputation for hardcore clubbing over the years, with Ios, Mykonos and most recently Paros leading the way. In Asia, Thailand's Ko Phangan is celebrated as the home of the Full Moon Party, the monthly rave on the beach that has attracted 10,000–20,000 people per event since the late 1980s. Nearby Ko Samui boasts an infamous Black Moon Party.

 # TV island shows

Islands have proved fertile ground for TV shows, with US classics such as Gilligan's Island (1964–67) and Fantasy Island (1978–84) – remember "Da plane, da plane"? At least nine seasons of the reality TV hit Survivor (US version) have been filmed on islands, while the popular current drama, Lost, takes place on a stereotypical desert/tropical island.

 # Prisons

Islands make excellent prisons. Napoleon Bonaparte was exiled first to Elba off the coast of Italy, then St Helena in the mid-Atlantic after Waterloo in 1815 – he died there in 1821. The most notorious prison in the US is Alcatraz, an island in San Francisco Bay that's now a tourist attraction. The same is true of Robben Island, off Cape Town, where Nelson Mandela was held captive for 18 years. Devil's Island in French Guiana, a French penal colony until 1946, was immortalized in the movie Papillon, while Tasmania started life (for Europeans at least) as an equally infamous prison colony in the nineteenth century. Asia is littered with former island prisons: Phu Quoc (Vietnam), Green Island (Taiwan), Ko Tarutao (Thailand) and the Andaman Islands (India).

Diving

▸▸ Top five islands for diving

Cozumel - Just off the Yucatan coast (Mexico), with an incredibly chromatic reef.

Little Cayman Island - Celebrated for the Bloody Bay Wall, a vertical drop that starts at around 6m before plunging down a sheer slope.

Palau - North Pacific island chain with World War II wrecks, stunning reefs and plenty of sharks.

Pulau Sipadan - Tiny coral island off the coast of Sabah, Malaysia, rising 600m off the sea bed and surrounded by steep drops on all sides that are populated by sharks, barracuda and giant parrotfish.

Truk - Officially known as Chuuk in Micronesia, with an impressive ensemble of sunken World War II Japanese battleships in the main lagoon.

Explorers

Christopher Columbus may have "discovered" the New World in 1492, but he didn't get anywhere near the mainland, landing at the island of San Salvador in the Bahamas and later Cuba and Hispaniola, establishing a colony in the latter. Viking explorers had founded a settlement on Newfoundland 500 years earlier. The Portuguese explorer **Ferdinand Magellan** led the earliest expedition to circumnavigate the globe, becoming the first European to call at Tierra del Fuego and the Philippines, where he died in 1521. And Dutch captain **Abel Tasman** was the first European to pull up to the shores of Tasmania and New Zealand on his voyage of 1642–44.

British navigator **Captain Cook** made three voyages around the Pacific between 1768 and 1779, becoming the first European to reach the Hawaiian islands, circumnavigating New Zealand and Newfoundland, and also visiting Tahiti, Easter Island and Indonesia. He discovered South Georgia and the South Sandwich Islands, but was killed in a fight with Hawaiians on the Big Island in 1779.

Charles Darwin's sea voyage in HMS *Beagle* (1831–36), vividly described in his *Journal and Remarks*, helped him develop his theory of evolution. On the voyage the ship called at numerous islands including Madeira, Cape Verde, Tierra del Fuego, the Falklands, Chiloé (Chile) and the Galápagos, where Darwin was to make some of his most important observations. On the return leg he stopped at Tahiti, New Zealand, Tasmania, Keeling (Cocos) Islands, Mauritius, St Helena, Ascension Island and the Azores.

> *"At sunset Martin Alonzo called out with great joy from his vessel that he saw land, and demanded of the Admiral a reward for his intelligence."*

**Christopher Columbus on the
discovery of San Salvador, Bahamas**

 # Desert island dramas

South Pacific (1958). Film version of the Rodgers and Hammerstein musical set on a South Pacific island during World War II.

Blue Lagoon (1980). Classic tale of two adolescents growing up together on a desert island, remake of earlier movies in 1923 and 1949.

Dr No (1962). In the first 007 flick, Bond is sent to steamy Jamaica to investigate the death of fellow operative, which leads him to the shadowy Dr No on Crab Key.

Pirates of the Caribbean (2003, 2006). Disney ride-turned-movie featuring pirates and rum-soaked islands in the sun.

Castaway (2000). Tom Hanks stars in an updated version of the Robinson Crusoe story.

 **Disasters**

The single greatest crisis faced by every island on the planet today is climate change, specifically rising sea levels, but they face many other natural and manmade disasters:

Tsunamis The South Asian Tsunami of 2004 devastated the Maldives, Sri Lanka, Sumatra and Phuket. The eruption of Krakatoa in 1883, off the coast of Sumatra, led to a similarly destructive wave.

Hurricanes Caribbean islands are frequently ravaged by hurricanes. The Great Hurricane of 1780 killed over 22,000 people in Barbados, Martinique and Sint Eustatius.

Earthquakes The Great Kanto earthquake of 1923 destroyed most of Tokyo and killed over 100,000 people, while Taiwan was badly hit by the Chi-Chi earthquake in 1999 which killed over 2000.

Extinctions Over-hunting by humans and the introduction of non-native animals has led to the extinction of many island species, most famously the Dodo of Mauritius, the giant Moa of New Zealand and the Tasmanian Tiger (Thylacine).

Pollution Nauru has been devastated by phosphate mining, while the Ok Tedi Copper Mine in Papua New Guinea is one of the world's biggest river polluters. Easter Island suffered ecological meltdown in the eighteenth century after deforestation and overpopulation.

 Bridges & causeways

Name	Total length
King Fahd Causeway – Bahrain to Saudi Arabia	26km
Penang Bridge – Penang to mainland Malaysia	13.5km
Great Seto Bridge – Shikoku to Honshu (Japan)	13.1km
Confederation Bridge – Prince Edward Island to mainland Canada	12.9km
Oresund Bridge – Zealand (Denmark) to Sweden	7.845km
Great Belt Fixed Link – Zealand to Funen (Denmark)	6.79km
Öland Bridge – Öland island to Kalmar (Sweden)	6.072km
Zeeland Bridge – Schouwen-Duiveland to Noord-Beveland (Netherlands)	5.022km

 **Food and drink**

Thousand Island Dressing, a blend of eggs, mayonnaise, ketchup, onions, pickles and bell peppers, is said to have been invented in the early twentieth century by Sophia LaLonde, the wife of a fishing guide who worked in the Thousand Island region of the Saint Lawrence River, between upstate New York and Ontario, Canada.

▸▸ Island drinks

Drink	Island	Chief ingredient
Arrack	Sri Lanka	coconut
Brennivín	Iceland	potato
Kaoliang liquor	Taiwan	sorghum
Kava	South Pacific islands	kava root
Madeira wine	Madeira	grapes
Marsala wine	Sicily	grapes
Rum	Barbados, Cuba, Puerto Rico	sugarcane
Zivania	Cyprus	grapes

 The price of paradise

Musha Cay in the Bahamas is the world's most expensive private resort island: it's a whopping US$3m per weekend to rent the whole thing. The celebrity hangout of Mustique is a comparative bargain, with its 85 luxury villas in the Grenadines going from US$4000 to US$35,000 each per week, depending on the season.

Language

The remoteness of some islands means that ancient languages can survive for much longer than expected. Papua New Guinea has just five million people but over 850 languages. Taiwan's aboriginal population, making up only 2% of the current population, once spoke at least 26 different languages, leading linguists to conclude that it is the home of the Austronesian language family which now covers most of the South Pacific and Southeast Asia.

"Foam will always find its way to the shore."

Solomon Island proverb

Spice Islands

The Spice Islands usually refer to the Maluku Islands in Indonesia, the original home of spices such as clove, mace and nutmeg. These were so valuable in Medieval and Early Modern Europe as crucial enhancements for food that European attempts to discover an alternative to the monopoly trade held by Venice was a driving factor in their voyages of discovery from the fifteenth century on. The Dutch eventually occupied the islands, controlling the trade until the British smuggled seeds out of the archipelago in the early nineteenth century.

Art

Paul Gauguin's love affair with island life began with Martinique in the1880s, followed by productive periods in Tahiti (1891–93 and 1895–1901), and the Marquises (1901–03). His post-impressionist paintings of "natural" Tahitian men and women are among the most popular from this period, balancing real life with an abstract ideal of nature.

Canaletto (1697–1768) painted beautiful landscapes of eighteenth-century Venice, the world's loveliest island city, documenting with obsessive accuracy the Grand Canal, Doge's Palace and everyday life on the canals.

Chen Cheng-po was one of Taiwan's greatest artists, with paintings such as Streets of Chiayi (1926) skillfully evoking the languid, tropical life on the southern half of the island in the early twentieth century.

Bali has a rich artistic tradition in carving, painting and sculpture, inspiration for European painters who were attracted to the island in the 1920s: **Walter Spies**, **Le Mayeur** and **Theo Meier** among them. Spies, in particular, created some mesmerizing images, blending the primitivism of Gauguin with Balinese myth in dark masterpieces such Calonarang (1930).

 Wildlife

▸▸ Unique island species

Giant Tortoise Galapagos Islands, Ecuador
Komodo dragon Lesser Sunda Islands, Indonesia
Kiwi New Zealand
Tasmanian Devil Tasmania, Australia
Philippine Tarsier Bohol, Samar, Leyte and Mindanao, Philippines
Lemurs Madagascar

*"A gentle breeze blowing in the right
direction is better than a pair of strong oars."*

Canary Islands proverb

21 Holy islands

Iona (563 AD) in Scotland and Lindisfarne (635 AD) off the northeast coast of England became popular monastic centres in the early years of British Christianity, though Anglesey in Wales was the holy island of the druids long before the arrival of the Romans. Apollo and Artemis are said to have been born on the holy Greek island of Delos. Meizhou Island, off the coast of Fujian, China, was the home of the popular goddess Mazu, and is now the site of the "mother" Mazu temple. Tana Kirkos in Lake Tana, Ethiopia, is inhabited by Christian monks who believe that it once contained the Ark of the Covenant, while remote Olkhon in Lake Baikal serves as the sacred island of the Buryat people.

"In the South Seas, the Creator seems to have laid himself out to show what He can do."

Rupert Brooke

22 Music

Numerous islands have produced unique – and enduring – musical genres, including:

Bali – gamelan music

Barbados – soca, Tuk, spouge

Cuba – salsa, son, mambo

Dominica – cadence

Dominican Republic – merengue, bachata

Fiji – folk music, meke

Guadeloupe/Martinique – zouk

Haiti – kompa, kadans

Hawaii – folk (mele, hula), hapa haole, slack-key guitar, Jamaiian, hip-hop

Jamaica – ska, reggae, rocksteady, ragga, dub

Puerto Rico – bomba, danza, salsa, reggaeton

Sri Lanka – baila

Trinidad – calypso, chutney, soca, parang

Zanzibar – kidumbak, ngoma, taarab

 # Surf's up

Hawaii – The home of surfing, with the mother of all breaks at Pipeline (Oahu's North Shore), and Waikiki, the surf beach that started it all.

Indonesia – Best surf in Southeast Asia at Bali (Kuta, Uluwatu, Nusa Dua), and G-Land (Bay of Grajagan, East Java).

Puerto Rico – Rincon has been a surf centre since hosting the World Championships in 1968.

Sri Lanka – Arugam Bay sees giant waves crashing in from Antarctica, one of the greatest breaks in Asia.

Tahiti – Teahupo'o and Papara Beach boast the most awesome surfing in the South Pacific.

 # The world's youngest island

Surtsey rose from the ocean bed off Iceland after a volcanic eruption in 1963. Since the eruption finished in 1967, erosion and subsidence have reduced its size, though it's unlikely that Surtsey will ever disappear entirely, as much of it is covered by hard lava flows, which are resistant to wind and water. Plants, such as mosses and lichen, and birds, including Atlantic Puffins, have colonized the island incredibly fast, but humans still need permission to land.

25 Atlantis

The Atlantis myth is one of the most famous of all "lost world" legends, an island state first mentioned by Plato. According to him Atlantis lay "beyond the pillars of Hercules", and eventually sank into the ocean around 9000 years before his birth. Speculation on just where exactly this place has focused on numerous locations, including North Bimini Road in the Bahamas, the Azores, Spartel Bank near the Straits of Gibraltar, and Crete, Sardinia and Santorini in the Mediterranean.

"How I wish that somewhere there existed an island for those who are wise and of good will."

Albert Einstein

25

Ultimate
experiences
Islands
small print

ROUGH GUIDES
give you the complete experience

"The complete guide for the thinking traveller...
enthusiastic and upbeat, providing a fascinating
cultural and political background"
The Observer reviewing *The Rough Guide to Cuba*

Broaden your horizons

www.roughguides.com

ROUGH GUIDES – don't just travel

We hope you've been inspired by the experiences in this book. There are 24 other books in the 25 Ultimate Experiences series, each conceived to whet your appetite for travel and for everything the world has to offer. As well as covering the globe, the 25s series also includes books on **Journeys, World Food, Adventure Travel, Places to Stay, Ethical Travel, Wildlife Adventures** and **Wonders of the World**.

When you start planning your trip, Rough Guides' new-look guides, maps and phrasebooks are the ultimate companions. For 25 years we've been refining what makes a good guidebook and we now include more colour photos and more information – on average 50% more pages – than any of our competitors. Just look for the sky-blue spines.

Rough Guides don't just travel – we also believe in getting the most out of life without a passport. Since the publication of the bestselling Rough Guides to **The Internet** and **World Music**, we've brought out a wide range of lively and authoritative guides on everything from **Climate Change** to **Hip-Hop**, from **MySpace** to **Film Noir** and from **The Brain** to **The Rolling Stones**.

Publishing information

Rough Guide 25 Ultimate experiences
Islands Published May 2007 by Rough
Guides Ltd, 80 Strand, London WC2R 0RL
345 Hudson St, 4th Floor,
New York, NY 10014, USA
14 Local Shopping Centre, Panchsheel
Park, New Delhi 110017, India
Distributed by the Penguin Group
Penguin Books Ltd,
80 Strand, London WC2R 0RL
Penguin Group (USA)
375 Hudson Street, NY 10014, USA
Penguin Group (Australia)
250 Camberwell Road, Camberwell,
Victoria 3124, Australia
Penguin Books Canada Ltd,
10 Alcorn Avenue, Toronto, Ontario,
Canada M4V 1E4
Penguin Group (NZ)
67 Apollo Drive, Mairangi Bay, Auckland
1310, New Zealand

Printed in China
© Rough Guides 2007
No part of this book may be reproduced
in any form without permission from the
publisher except for the quotation of brief
passages in reviews.
80pp
A catalogue record for this book is
available from the British Library
ISBN: 978-1-84353-832-6
The publishers and authors have done
their best to ensure the accuracy
and currency of all the information in
Rough Guide 25 Ultimate experiences
Islands, however, they can accept
no responsibility for any loss, injury, or
inconvenience sustained by any traveller
as a result of information or advice
contained in the guide.
1   3   5   7   9   8   6   4   2

Rough Guide credits

Editor: AnneLise Sorensen
Design & picture research: Jess Carter
Cartography: Katie Lloyd-Jones,
Maxine Repath

Cover design: Diana Jarvis, Chloë Roberts
Production: Aimee Hampson,
Katherine Owers
Proofreader: Anna Owens

The authors

Chris Hamilton (Experience 1) guides educational sailing adventures in the Caribbean and contributes to the Rough Guide to Antigua & Barbuda. Keith Drew (Experience 2) is a Rough Guides editor and has travelled throughout Central America. Mark Ellwood (Experience 3) writes the Rough Guide to Miami & South Florida. Donald Reid (Experience 4) writes the Rough Guide to Scotland. Phil Lee (Experience 5) is author of the Rough Guide to Norway. Jens Finke (Experience 6) is author of Rough Guides to Tanzania and Zanzibar. Polly Thomas (Experience 7) is co-author of Rough Guides to Trinidad & Tobago and Jamaica. Robert Andrews (Experience 8) writes the Rough Guide to Sicily. David Dalton (Experience 9) writes the Rough Guide to the Philippines. AnneLise Sorensen (Experience 10) grew up visiting family in Jutland and covers Denmark for guidebooks and magazines. Henry Stedman (Experience 11) is co-author of the Rough Guide to Indonesia. Mark Ellingham (Experience 12), Rough Guides founder and series editor, travels regularly to the Scilly Isles. David Leffman (Experience 13, 15, 22) is co-author of the Rough Guide to Iceland. Christian Williams (Experience 14) writes the Rough Guide to Tenerife & La Gomera. Matt Norman (Experience 16) is co-author of the Rough Guide to Cuba. Greg Ward (Experience 17) writes the Rough Guide to Hawaii. Simon Richmond (Experience 18) is co-author of the Rough Guide to Japan. John Fisher (Experience 19) writes Rough Guides to Crete and Greece. Gavin Thomas (Experience 20) writes the Rough Guide to Sri Lanka.
Sean Harvey (Experience 21) writes the Rough Guide to the Dominican Republic. Richard Trillo (Experience 23) writes the Rough Guide to West Africa. Stephen Keeling (Experience 24, Miscellany) has travelled to more than fifty islands, and has sampled kava in Fiji, Vanuatu and New Caledonia. Sarah Hull (Experience 25) contributes to the Rough Guide to New England.

Picture credits

Over 70 reference books and hundreds of travel guides, maps & phrasebooks that cover the world

ROUGH GUIDES

Australia
Cuba
Britain
Singapore
Vietnam
New York City

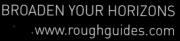
BROADEN YOUR HORIZONS
www.roughguides.com

Index